Materialien für den bilingualen Unterricht
Geographie

Polar Zones

Places of Harshness

Cornelsen

Materialien für den bilingualen Unterricht: Geographie
Polar Zones – Places of Harshness

Mit Beiträgen von
Joan Fugel, Ricarda Hampel, Anne Reif (in Zusammenarbeit mit der Verlagsredaktion)

Das Modul wurde entwickelt auf der Grundlage eines Kapitels im bilingualen Lehrwerk *Around the World 1 (New Edition)* (ISBN 978-464-31182-0).

Redaktion: Dr. Frank Erzner, Marc Proulx
Redaktionsassistenz: Merlene Griffin
Karten und Grafik: Dieter Stade, Hemmingen; Peter Kast, Ingenieurbüro für Kartografie, Schwerin
Illustrationen: Hans Wunderlich, Berlin; Volkmar Rinke, Hildesheim
Umschlaggestaltung: Klein & Halm Grafikdesign, Berlin
Technische Umsetzung: Arnold und Domnick, Verlagsproduktion – alle Medien, Leipzig

www.cornelsen.de

1. Auflage, 1. Druck 2009

Alle Drucke dieser Auflage sind inhaltlich unverändert und können im Unterricht nebeneinander verwendet werden.

Druck: CS-Druck, Cornelsen-Stürtz, Berlin

ISBN: 978-3-06-061884-2

Inhalt gedruckt auf säurefreiem Papier aus nachhaltiger Forstwirtschaft.

Preface

Texts
which give you an overview of or introduce you to the subject.

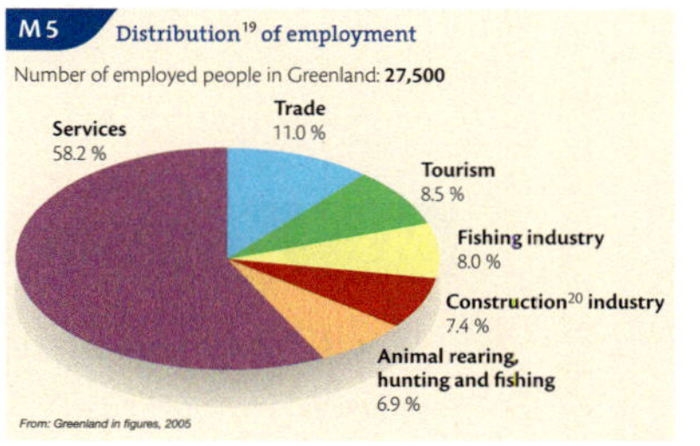

Materials boxes
which present statistics, diagrams, and photos for you to work with. They are labelled with an 'M' and numbered throughout the module.

Key terms

South Pole, Antarctica, Cold Pole, inland ice, shelf ice, pack ice, Arctic, Arctic Circle, iceberg, North Pole, tundra, permafrost, Inuit, Greenland, Gulf Stream, Siberia, coniferous/deciduous forest, taiga, Gazprom, Lukoil, Trans-Alaska Pipeline

Key terms
These are important terms used in the module which are marked in bold. These terms are useful for your revision of the topic. You will find all the key terms listed in the 'Key terms' box at the end of the module.

New vocabulary
All numbered words[1] are translated or explained on the same page.

Tasks
In each section you will find different kinds of tasks which help you to understand a phenomenon or issue, for example

- by **organizing information** in charts and diagrams to simplify and summarize information, making it easier to understand and remember.
- by **making connections**: tasks in which you think critically about problems, draw comparisons or think about the effect that certain geographical issues have on your life.
- by **doing extra research**: tasks that tell you to research a topic further, either on the Internet or using other reference material.

Activate your English

Asking questions in an interview

Good morning/afternoon/... /Hello ...
I would like to ask you / May I ask you ... some questions about ...(?)
Where do you ... / How do you ... / How many/much ...?
What problems are there ... / How has life changed in/for ...?
Do you think ... better/worse ... / What are the (dis)advantages of ...?

Activate your English
These boxes provide you with words and phrases for writing and speaking about the topics which are dealt with in the tasks.

Revision and further exercises
At the end of the module there are activities which help you revise what you have already learned and others which make you think further about the main issues of the module.

Skills pages
The geographical skills found on pp. 18–21 give you strategies and tools for working with the tasks and the different kinds of information you will deal with in the module.

Appendix
In the Appendix you will find an English-German vocabulary list with translations and pronunciation of important terms used in the sections and skills pages; a geographical dictionary (German-English), which will help you find words for use in your tasks and discussions; and a list of the abbreviations and symbols used throughout the module.

1 Journeys to the end of the earth

M 1 Race to the South Pole

*This is a very bad day. We are all tired and have cold feet and hands. It is –30 ° Centigrade and there is a snowstorm. ... There is food for four men, not five. ... Oates died like a good Englishman. We all did (!) Please remember us and look after our families. We did our best. ...
Great God this is an awful place.*

It is January 1912 and the Englishman Sir Robert Scott is writing to his wife at home in England. He and his men have fought their way to the **South Pole**[1] only to find that the Norwegian Roald Amundsen has got there before them.

M 2 Amundsen at the South Pole

M 3 Scott and his men

Scott and his men never reached the ship that was waiting to take them home. They died only 18 km from the last depot[2] where there was shelter[3] and supplies[4] which would have saved their lives.

1 South Pole *Südpol*
2 depot *Lager*
3 shelter *Schutz, Obdach*
4 supply *Vorrat*

M 4 Amundsen's and Scott's routes

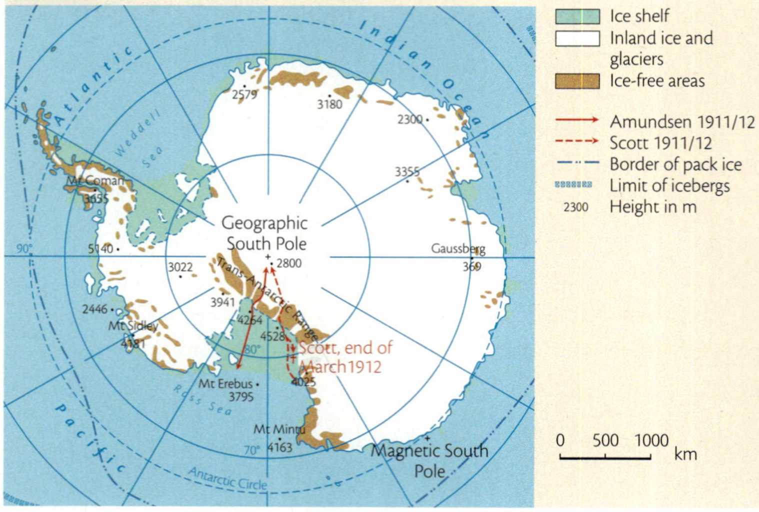

M 5 Satellite image of Antarctica

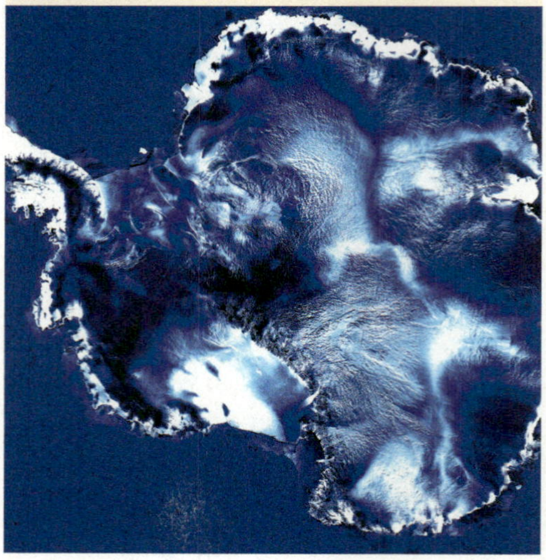

Explorers[5] of today

M 6 Messner and Fuchs ...

M 7 ... at the South Pole

Here are two modern explorers – Arved Fuchs and Reinhold Messner – standing at the South Pole. Look at their clothing and the tent they slept in.

M 8 **This is what Messner wrote in his book about his journey:**

In our tent we were quite comfortable. In the evenings we sat around the cooker, melted snow and used the water to make tea and soup. After that we had bacon and hard bread, then rice, potatoes or noodles with meat. For breakfast we had coffee, bacon, bread and biscuits. The meals usually lasted about two hours. Before we moved on we filled our thermos flasks with vitamin and mineral drinks. Then we took down the tent, put everything on our sledges[6], then put on our skis and the nylon harnesses[7] that we used to pull the sledges – just like horses. The route down from the Beardmore Glacier[8] was especially difficult. There was nothing but ice for 150 km. One time I was a bit ahead of my partner Arved when I was suddenly surrounded by lots and lots of crevasses[9]. I couldn't pull my sledge any further. I tried but I had to give up because some of the crevasses were as big as a church. I walked back and after a time found a way out. At the bottom of a huge ice hole about 200 m deep, we finally put up our tent.

5 explorer *Entdecker*
6 sledge *Schlitten*
7 harness *Geschirr*
8 glacier *Gletscher*
9 crevasse *Gletscherspalte*
10 equipment *Ausrüstung*

1. Using the information in M 4 and a map of **Antarctica** in your atlas, follow and describe the routes that Scott and Amundsen took. Discuss what difficulties they must have had.
2. Why did both men choose the months of October to December for their expeditions?
3. Write an earlier entry in Scott's diary.
4. Do you think Fuchs's and Messner's journey was easier or more difficult than that of Scott or Amundsen?
5. With a partner or in groups, make a list of the equipment[10] you would need and discuss how you would prepare such an expedition today.

Activate your English

Expressing opinions
I (don't) think (that) ...
In my opinion, ...
It seems to me that ...
On the one hand, ... On the other, ...
That's right/correct/true (but) ...
I (don't) agree / I disagree, because ...

Discussing preparations
If I/you were doing/going on such an expedition today, ...
I/You would (probably) need/have to take along ...
To prepare for the expedition I would train/exercise/research/...
I would prepare by first of all researching ...

2 Antarctica and the Arctic – A comparison

Antarctica – A continent surrounded by[1] sea

M 1 Profile (cross-section) of Antarctica

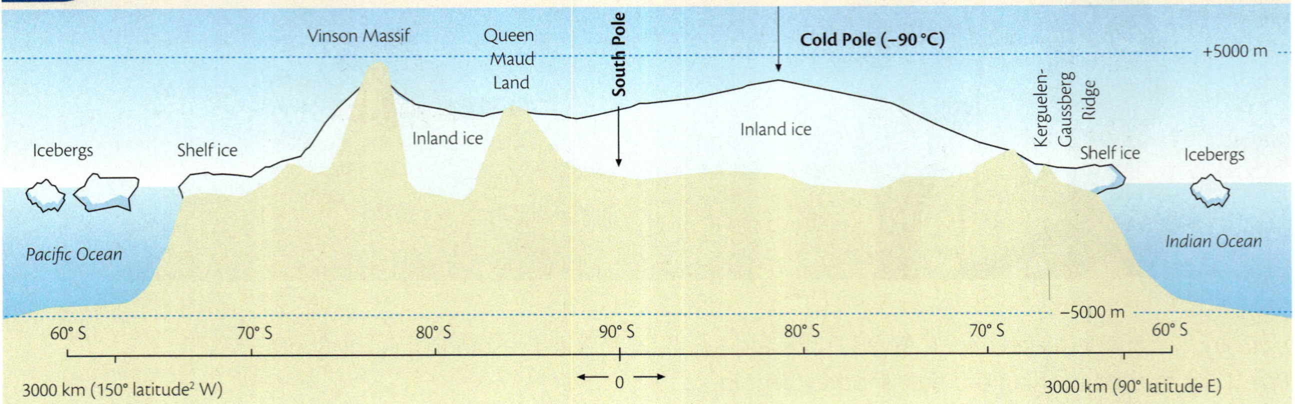

Did you know?

Antarctica ...
- is the coldest and the highest continent*
- is nearly twice as big as Australia
- is a cold desert – it snows very little in the interior[3]
- 80–90 % of the world's fresh water[4] locked in its ice
- has the thickest ice sheet[5] – more than 4000 m (4 km!) deep
- has the strongest winds – more than 200 kmh[6] at times
- is rich in marine life[7] and mineral resources[8].

Average inland temp. -57 °C; lowest recorded temp. -89.6 °C in Vostok; average height 2300 m; highest peak[9] Vinson Massif, 5140 m

M 2 Home of the penguin

1 surrounded by *umgeben von*
2 (line of) latitude *Breitenkreis*
3 interior *Landesinneres*
4 fresh water *Süßwasser*
5 ice sheet *Eisdecke*
6 kmh = kilometres per hour *Kilometer pro Stunde*
7 marine life *Meerestiere*
8 resource *Bodenschatz*
9 peak *Bergspitze*
10 prove *beweisen*
11 above sea level *über dem Meeresspiegel*
12 Cold Pole *Kältepol*
13 inland ice *Inlandeis*
14 shelf ice *Eisschelf*
15 pack ice *Packeis*
16 plot a graph *ein Diagramm zeichnen*
17 line graph *Liniendiagramm*

1 Find a map of Antarctica in your atlas. Together with a partner, use the information to do the following tasks:

a How many of the above statements can you prove[10] to be true? Now look at the symbols in the key to your map, then take a closer look and answer the following questions:

b How deep is the ice sheet at its deepest point and how high is the ice at its highest point above sea level[11]?

c Do any people live there? If so, why?

2 Find out what animals live in the seas around the continent.

3 Explain the terms: **Cold Pole**[12], 'geographical South Pole', **inland ice**[13], **shelf ice**[14] and **pack ice**[15].

4 Look at this set of climate data from two polar stations. Plot[16] two line graphs[17] for the temperatures.

J	F	M	A	M	J	J	A	S	O	N	D
-27 °C	-28 °C	-26 °C	-19 °C	-6 °C	2 °C	4 °C	4 °C	-1 °C	-9 °C	-18 °C	-24 °C
-3 °C	-9 °C	-18 °C	-23 °C	-23 °C	-25 °C	-27 °C	-29 °C	-23 °C	8 °C	6 °C	7 °C

→ **Skills**, 'Reading and plotting climate graphs', pp. 18-19

5 Describe the two graphs and say which station is McMurdo in Antarctica and which is Barrow in the Arctic. Explain your choice.

The Arctic – An ocean surrounded by land

The **Arctic** is the name we give to the areas within the **Arctic Circle** (66.5° N). At the centre is the Arctic Ocean, which is five times as big as the Mediterranean Sea[18] and is frozen over in winter.

M 3 Profile (cross-section) of the Arctic

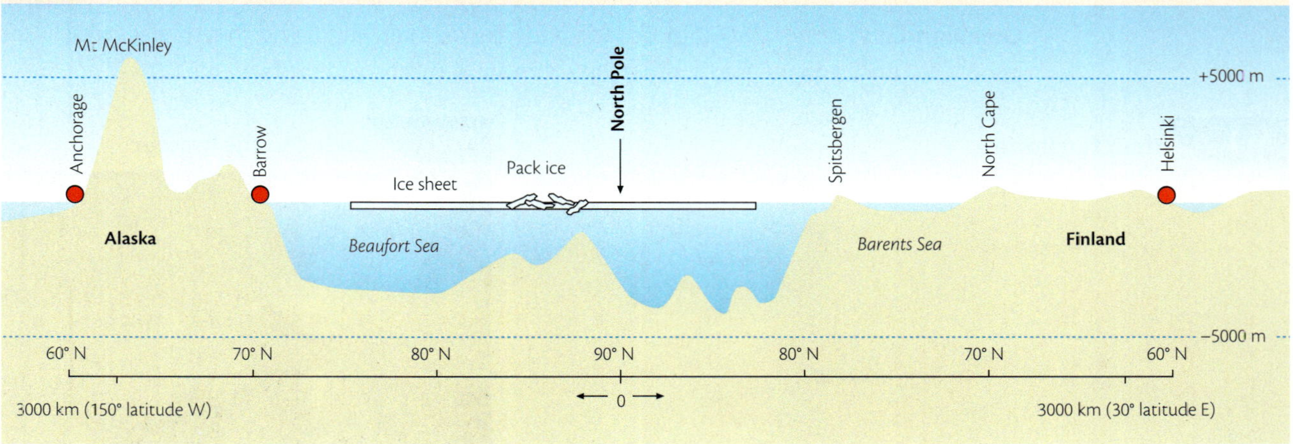

In summer the ice begins to melt and turn into pack ice, which is moved by winds and tides[19]. Some of it drifts[20] south on ocean cur rents[21] as **icebergs,** which are dangerous to ships. There is no land at the **North Pole**.

The vegetation[22] of the lands of the far north is called **tundra**,[23] which is a flat, treeless[24] area. It is too cold for trees to grow here. Just below the surface,[25] the ground is permanently frozen (sometimes to a depth[26] of 600 metres). We call this **permafrost**[27].

In the summer the top layer[28] melts, but the water cannot drain away[29] because of the frozen soil underneath, so the ground becomes wet and marshy[30]. This ground is difficult to build houses and roads on.

M 4 A polar bear on the Arctic ice

Although there are no trees, other plants grow here: small shrubs[31], heathers[32], mosses[33] and different kinds of lichens[34], which can all stand being buried under the snow in the long Arctic winter. Reindeer moss, which is a kind of lichen, is the main food of the caribou[35] (tamed caribou are called 'reindeer' in Europe). These animals move southwards in winter looking for new grazing grounds.

Other land animals also manage to live in this harsh[36] climate such as musk oxen[37], wolves, lemmings and Arctic foxes, which grow a thick white fur[38] in winter to keep themselves warm (and safe). Polar bears[39] are the most dangerous animals of the Arctic and live mostly on fish, seals[40] and birds. They are extremely good swimmers and have been seen 300 miles from land. Unfortunately some of them have learned to look for food in towns and villages and have become dangerous to the people living there, so they have to be chased away[41] or even shot.

6 Use the information on these two pages and in your atlas to make a table comparing the two polar regions of the earth. Use the following headings: size, countries, landscapes, vegetation, climate, animals, population/settlements[42].

18 Mediterranean Sea *Mittelmeer*
19 tide *Gezeitenstrom*
20 drift *treiben*
21 ocean current *Meeresströmung*
22 vegetation *Vegetation (pflanzlicher Bewuchs)*
23 tundra *Tundra*
24 treeless *baumlos*
25 surface *Oberfläche*
26 depth *Tiefe*
27 permafrost *Dauerfrostboden*
28 layer *Schicht*
29 drain away *wegsickern, ablaufen*
30 marshy *sumpfig*
31 shrub *Strauch*
32 heather *Heidekraut*
33 moss *Moos*
34 lichen *Flechte*
35 tamed caribou *gezähmtes Karibou*
36 harsh *rau*
37 musk ox(en) *Moschusochse(n)*
38 fur *Pelz*
39 polar bear *Eisbär*
40 seal *Robbe*
41 chase sb./sth. away *jdn./etwas wegjagen*
42 settlement *Siedlung*

3 The changing face of the Arctic

An Arctic culture then and now

Life north of the Arctic Circle is far from easy and yet the **Inuit**[1] people live there – 45,000 in **Greenland** and about 70,000 in Canada and Alaska. Inuit, the name they prefer, means 'the people'. Eskimo, a name used for them in some areas, means 'eaters of raw meat[2]'.

M 1 Inuit territories

M 2 Living in the 'freezer'

Nunavut ('our land') in the north of Canada, the first area ruled entirely by the Inuit, is six times as big as Germany but has only 26,000 people. This is not surprising, as temperatures fall to −40 °C in winter and only rise[3] to about 10 °C in summer.

We often think of the Inuit as primitive peoples[4] living in igloos and hunting seals and whales[5]. This traditional way of life, however, is disappearing quickly. Nowadays most Inuit live in heated wooden houses, buy their food in well-stocked[6] supermarkets, go to school, travel around in skidoos (snowmobiles) and communicate via mobile phones and computers. Most are bilingual and sometimes find living between two languages and two worlds quite difficult.

M 3 Living in an igloo

M 4 Modern housing

M 5 An older Inuit remembers:

We lived together as a large family, my parents, my wife and children, and my brothers with their wives and children. We provided[7] almost everything we needed ourselves. In the summer we hunted seals from our kayaks with harpoons, later with guns. In the winter we found the seals at their breathing holes in the ice. Sometimes we were lucky and killed a caribou that was swimming. ... We used umiaks to take our tents to the best fishing grounds and to go whale hunting. We used the train oil[8] from whales and seals for the lamps and stoves[9] in our winter huts. When we went hunting in the winter, we built igloos from blocks of snow and ice. I remember one year when there were no seals. We suffered very badly.

1 Inuit (sing) *Inuk*
2 raw meat *rohes Fleisch*
3 rise *steigen*
4 peoples *Völker*
5 whale *Wal*
6 well-stocked *gut gefüllt, ausgestattet*
7 provide sth. *etwas herstellen*
8 train oil *Lebertran*
9 stove *Herd, Kochstelle*

M6 Isavarag, 40 years old, works on a fish trawler[10]

Work on board the ship is difficult. We are often away for days and days. I have two children, 12 and 16 years old. My wife works now and then in the supermarket.
Margie is in her last year at school. She is a good pupil and would like to be a nurse.
Georgie likes watching TV. At school he is more interested in the new subjects – hunting, fishing and animal furs[11].
If I have time at the weekend, we take the dog sledges and go hunting. Thirty years ago my family lived in a wooden hut on the coast[12]. Mother used to sew[13] anoraks and boots for us from sealskins. At bedtime grandma told us stories about courageous hunters.
Today seals are hunted with rifles[14]. We bring the dead animals on skidoos to a collection point. We can buy our clothes off the hook[15]. Instead of kayaks, hunters use boats made of polyester[16] with outboard engines. We can buy everything we need at the supermarket. We only go hunting for fun today. However, a lot of families can't get on without support[17] from the state. We are entitled to[18] a supplement[19] for children and grandpa gets a pension[20]. Sadly, more and more people are becoming dependent on unemployment benefits[21]. Our neighbour couldn't cope[22] with this situation and took to drinking.

M7 Traditional and modern ways of life among the Inuit

1 Using the notes below, make a table showing the Inuit way of life in the past and today.
– helicopters and planes as means of transport – hunting with harpoons from kayaks and umiaks – working for an oil company – vitamins from eating the raw meat of animals – hunting caribou, seals, polar bears and whales – clothes and tents made from the hides[23] of animals – electricity, TV and computers – tooth decay[24] as a result of poor nutrition[25] (sugary foods) – money used for trading[26] – tools[27] and weapons[28] made from the bones of animals – mining[29] of valuable resources such as oil … .

2 Study the two personal accounts[30] and the pictures on these pages and add more notes to your table.

3 Use these notes and other information you have to write an interview with
a a young Inuk of your age group,
b his or her grandmother.

4 Act out your interviews in groups or with a partner in class.

Activate your English

Asking questions in an interview

Good morning/afternoon/… /Hello …
I would like to ask you / May I ask you … some questions about …(?)
Where do you … / How do you … / How many/much …?
What problems are there … / How has life changed in/for …?
Do you think … better/worse … / What are the (dis)advantages of …?

10 trawler *Fischerboot*
11 animal fur *Tierpelz*
12 coast *Küste*
13 sew sth. *etwas nähen*
14 rifle *Gewehr*
15 off the hook *fertig genäht*
16 polyester *Polyester*
17 support *Unterstützung*
18 be entitled to sth. *auf etwas Anrecht haben*
19 supplement *Zuschuss*
20 pension *Rente*
21 unemployment benefits (pl) *Arbeitslosengelder*
22 cope with sth. *mit etwas fertig werden*
23 hide *Haut*
24 tooth decay *Zahnfäule*
25 nutrition *Ernährung*
26 trading *Handel*
27 tool *Werkzeug*
28 weapon *Waffe*
29 mining *Bergbau*
30 account *Bericht, Beschreibung*

4 Greenland

With an area of about 2.2 million km^2 (square kilometres) Greenland is the largest island on earth (Australia and Antarctica are considered continents). It is almost completely covered in ice, under which there are plateaus[1] and mountains, the mountain peaks sometimes sticking up like black fingers out of the white snowfields.

M 1 Greenland in winter

M 2 Greenland in summer

The mean monthly temperature[2] in Greenland never rises above 0 °C and the mean annual[3] temperature is –31 °C. A temperature of –65 °C has even been recorded there. It also snows in every month of the year, but as the temperature seldom rises above freezing point, very little of the snow melts, so it gets deeper and deeper and turns into ice. The sheet of ice which is formed can be over 3000 metres thick in some parts of the island's interior.

The ice sheet moves very slowly, pushed out from the centre by new layers of snow and ice and slides[4] towards the coast. There it is forced through gaps in the mountains to form glaciers which break off or 'calve'[5] when they reach the warmer waters of the sea. Sometimes the ice is still about 1000 metres thick at the coast, forming very high cliffs. Ocean currents carry the enormous icebergs away southwards, where they gradually melt as temperatures rise. It was one such iceberg that the passenger ship Titanic ran into on the night of 14th April, 1912, causing the ship to sink with the loss of many lives.

1 plateau *Hochebene*
2 mean monthly temperature *mittlere Monatstemperatur*
3 annual *jährlich*
4 slide *gleiten, rutschen*
5 calve sth. *etwas kälben; (hier) etwas abbrechen*
6 inlet *Meeresarm*
7 fjord *Fjord*
8 habitable *bewohnbar*

M 3 North Atlantic ocean currents

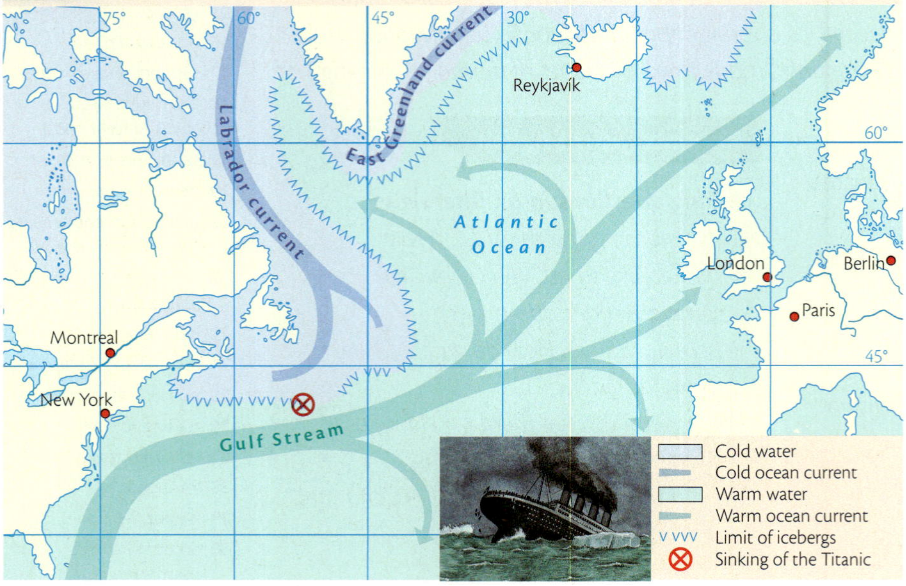

The coast of Greenland is rocky with many inlets[6] or fjords[7]. About 16 per cent of the land here is ice-free and habitable[8]. Most of the people who live there are Inuit, who first settled in the area about 3000 years ago. Danish whale hunters came in the 17th century and so many Danish settled there that Greenland became a province of Denmark in 1953. That is why the Danish language is spoken there, although Greenlandic has become the official language and since 1979 the island has had its own government.

Living on the fringe[9]

In the south-west of Greenland the 'green fringe' is the result of the warm **Gulf Stream**[10] or North Atlantic Drift[11], which flows[12] northwards from the Gulf of Mexico. Along this coast you can find areas with tundra vegetation in the form of grass, moss, heather and lichen. The larger settlements are found here and this is where valuable minerals[13] such as gold, platinum[14], aluminium ore[15] and uranium[16], which is used in nuclear power stations, have been discovered. The capital city is Nuuk, where there is often not enough housing for the people who leave the smaller settlements. The only roads are in the larger towns, otherwise people travel by ship, plane or helicopter. Housing is modern and up to European standards. The Internet and mobile phones are a normal part of everyday life. There are a lot of sports opportunities for young people – football is very popular in Greenland. Many young people also try to keep old traditions alive and spend time learning to build and handle kayaks.

The people here still live mainly from fishing, although this is not always easy as the harbours are sometimes blocked by pack ice even in spring. Also, because of over-fishing there are not as many fish as in the past, and the European Union[17] controls the rights of fishing and gives subsidies[18] to fishermen. Today tourism is becoming more and more important, as is also the mining of minerals.

9 fringe *Rand*
10 Gulf Stream *Golfstrom*
11 North Atlantic Drift *Golfstrom*
12 flow *fließen*
13 valuable minerals *wertvolle Mineralien*
14 platinum *Platin*
15 aluminium ore *Aluminiumerz*
16 uranium *Uran*
17 European Union *Europäische Union*
18 subsidy *Subvention*
19 distribution *Verteilung*
20 construction *Anlagenbau*
21 shore *Küste, Ufer*
22 densely populated *dicht besiedelt*

M 4 **A fish factory in Nuuk**

M 5 **Distribution[19] of employment**

Number of employed people in Greenland: **27,500**

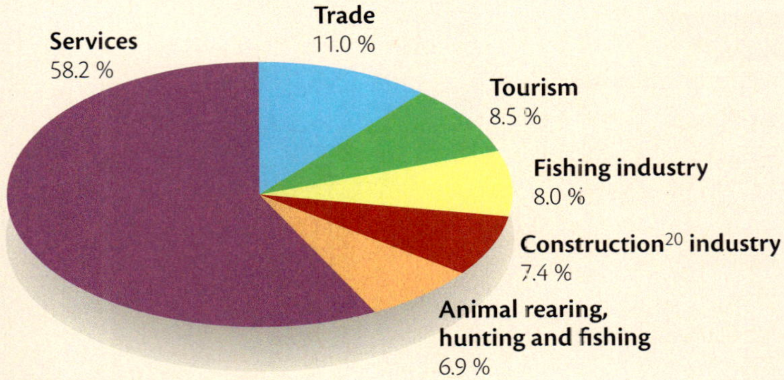

From: Greenland in figures, 2005

1 Look at M 3. Which ocean currents reach the shores[21] of Greenland?
2 Find a map of Greenland in your atlas. Which parts of Greenland are more densely populated[22]? Name the biggest settlements and describe their location.
3 Look at the pie chart in M 5 above. Describe and explain the distribution of employment.
4 Why does the Greenlandic government hope to depend less on fishing in the future?
5 With a partner or in a group, prepare two pages of information (text and pictures) about Greenland for a tourist brochure.

Activate your English

Describing geographical locations and features

... has an area of ... square kilometres
... is located in the north/south-west/... of the island
... south/north-east/... of Nuuk/...
... roughly/about ... kilometres from ...
... (in) the island's interior/inland area ...
... (along) the coastal fringe ...
... there is/are a sheet of ice / plateaus / mountains (peaks) / glaciers / lowland areas / a coastal plain ...
The coast is rocky with many inlets or fjords ...
There is very little vegetation / Vegetation is sparse ...
tundra vegetation / grass / moss / heather / lichen
There are valuable minerals such as ...
Valuable minerals can be found ...

5 Siberia – Icy treasure trove[1] of the North?

M 1 Cold Pole of the North

Oimjakon 63° 16′ N/143° 09′E
T -15,7 °C 740 m P 198 mm

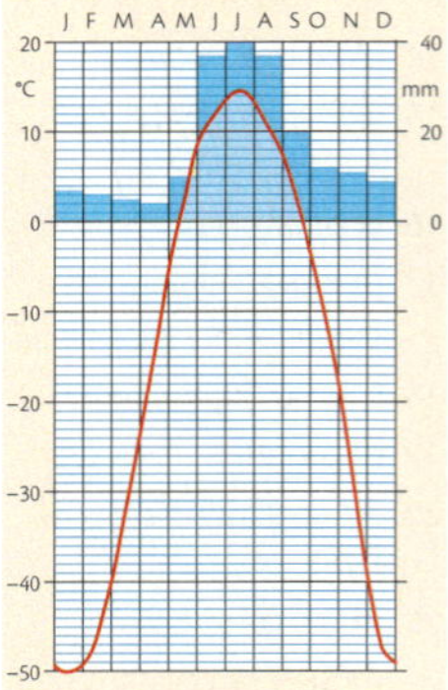

Siberia lies in the northern part of Russia and covers[2] an area much bigger than Europe. It is a region of dense[3] **coniferous forests[4] (taiga)** and treeless frozen wastes[5] (tundra). In this icy desert, where only 13 cm of rain and snow fall each year, summers are short and winters are bitterly cold and long. The Cold Pole of the northern hemisphere is here and temperatures can fall to –78 °C in winter.

In summer the frozen rivers, which are used as communication[6] and transport routes in winter, begin to thaw,[7] the ground becomes soft and muddy and there is extensive flooding[8].

M2 Landscapes of Siberia

M 3 Vegetation zones in Russia

- Bog[9]/moor
- Area of periodic drought[10]
- Tundra
- Taiga
- Deciduous[11] and mixed forest
- Steppes (grassland)
- Semi-desert[12] and desert
- Mountain vegetation

0 500 1000 km

M4 Why would anyone want to live here?

In fact not many people do. Not until the 16th century when the Kosaks crossed the Ural mountains did people begin to settle here. Today only ten per cent of the Russian population lives here, although many valuable resources such as coal, iron ore[13], oil, natural gas and even gold and diamonds lie under the frozen surface.

1 treasure trove *Schatz, Fundgrube*
2 cover sth. *etwas bedecken*
3 dense *dicht*
4 coniferous forest *Nadelwald*
5 waste *Ödland*
6 communication route *Verbindungsweg*
7 thaw *auftauen*
8 extensive flooding *weitläufige Überflutung*
9 bog *Sumpf*
10 periodic drought *wiederkehrende Dürreperioden*
11 deciduous forest *Laubwald*
12 semi-desert *Halbwüste*
13 iron ore *Eisenerz*

Norilsk – A fading light in the polar night?

In 1928 the Russian government began to mine some of the minerals in Siberia. At first political prisoners[14] were sent to work in the mines. Thousands died in the terrible conditions and the cold. Later people were offered high wages[15] and longer holidays to go and work there and new towns were built. One such town is Norilsk.

In the 1930s copper[16], nickel[17], coal[18] and iron ore were discovered near Norilsk. In the following years the 'Kombinat', as the industrial complex was called in communist times, took care of the workers and their families by building schools, hospitals and canteens. However, after the end of the Soviet Union in 1991, when there was no more support from the state, industry fell into decline[19] and the area around the city was left heavily polluted[20]. Thousands lost their jobs, along with their dreams of saving enough money to be able to move away to a better life further south.

There is, perhaps, a future for Norilsk – a nickel smelter[21] has begun to work again and the oil and gas which lie under the frozen surface may once again bring wealth to the region. There are also hopeful signs that in the future something may be done about the ecological disaster[22].

14 political prisoner *politischer Gefangener*
15 wage *Lohn*
16 copper *Kupfer*
17 nickel *Nickel*
18 coal *Kohle*
19 fall into decline *verfallen, rückgängig werden*
20 polluted *verschmutzt*
21 nickel smelter *Nickelverhüttungsanlage*
22 ecological disaster *ökologisches Desaster, Umweltkatastrophe*
23 smelting *Verhüttung*
24 non-ferrous metal *Buntmetall*
25 agriculture *Landwirtschaft*

M5 Street scene in Norilsk

1. Find Oimjakon in your atlas. Interpret the climate graph in M1.
2. Describe the pictures in M2 and match them with three of the vegetation zones in M3.
3. Make notes on the climate and landscapes of Siberia and explain why the conditions cause great difficulties for agriculture[25] and the mining and transport of mineral resources.
4. Give a detailed description of the map in M6 and explain the reasons for
   a the growth of Norilsk and
   b the difficulties of living and working there.
5. Role play: You and your family have decided to leave Norilsk. Your relatives and friends would like you to stay. Explain why you have decided to make this move. Write the dialogue and perform it in class.

M6 Norilsk: industrial island in the frozen tundra

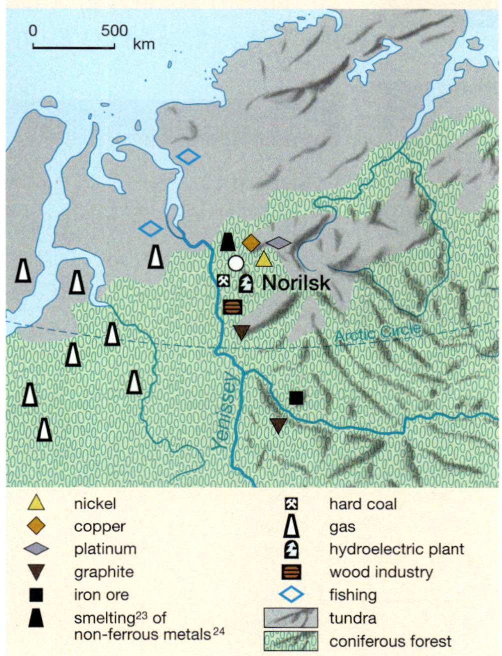

Activate your English

Describing thematic maps

Norilsk is located … / … in the far north of … / … north of the Arctic Circle …
It can be reached in summer/winter by …
The Yenissey flows from … to …
Vegetation consists of …
Agriculture is possible south of … because …
The surrounding area is rich in …

M7 Norilsk – some facts:

- Northernmost city of the world
- Population >174,000 (1989) <135,000 (2002)
- Next big city 2000 km
- Nearest railway station 1000 km
- Snow 250 days a year
- Polar night
- Permafrost
- Marshy swamps in summer (mosquitoes)
- River Yenissey frozen October to March

→ **Skills**, 'Reading and plotting climate graphs', pp. 18–19; 'Working with maps', pp. 20–21

6 Pipelines for the world

M 1 Siberian oil and gas

Natural gas pipeline

Oilfield in West Siberia

Scars in the landscape

Under the frozen earth of the tundra in West Siberia lie an estimated[1] 70 per cent of Russia's natural gas reserves[2] and 30 per cent of its crude oil[3].

Since the 1960s, 150 gas and oil fields with roads, railway lines, airfields and heliports have been developed[4].

Drilling[5] for oil and gas and laying pipelines in the frozen permafrost is an almost impossible task. Conditions in the long, dark winters are extremely difficult. It is so cold that the workers can only be outside for an hour at a time. If it is windy, work has to stop. In summer it is not much better as the ground gets too wet and soggy[6] to lay the pipelines.

Oil and gas are transported thousands of kilometres through pipelines, much of it to Western and Central Europe. The companies Nordlicht and Jamal-Europe bring oil and gas to us, and a new pipeline that runs through the Baltic Sea[7] to Greifswald in Germany has been planned for 2010 (see M 2).

Drilling for oil and gas in the tundra causes enormous ecological[8] problems. In the last 20 years the Russian government[9] has not been willing or able to spend money on modern technology and conservation[10] programmes. Rusting[11], leaking[12] pipelines and scars[13] in the landscape are the sad result of this neglect[14].

Key terms

Gazprom
The largest gas company in the world. Gazprom's network of pipelines would reach three times around the world.

Lukoil
The largest oil company in Russia, with a global network.

M 2 Oil and gas for Europe

1 estimated *geschätzt*
2 natural gas reserves *Erdgasvorräte*
3 crude oil *Rohöl*
4 develop sth. *etwas entwickeln*
5 drill *bohren*
6 soggy *matschig*
7 Baltic Sea *Ostsee*
8 ecological *ökologisch, Umwelt-*
9 government *Regierung*
10 conservation *Erhaltung der Umwelt*
11 rust *rosten*
12 leak *lecken, auslaufen*
13 scar *Narbe*
14 neglect *Vernachlässigung*

TAP – Trans-Alaska Pipeline

Great quantities of oil have also been found in the north of Alaska around Prudhoe Bay. The 'black gold', as it is sometimes called, makes up 85 per cent of the income[15] of Alaska and 25 per cent of the USA's oil. As the sea is frozen for most of the year, oil cannot be transported by this route. A 1200 km pipeline had to be built to transport the oil to the Valdez oil terminal[16] on Alaska's south coast. From there it is transported by oil tankers. Apart from the difficulties of laying a pipeline in the permafrost, other problems were: earthquakes[17]; the protests of the Inuit population who did not want a pipeline in their areas; the protection[18] of one of the last wildernesses[19] of the world; and concerns[20] about the migration[21] of caribou, whose trek south to look for food in winter would have been stopped by an overland pipeline. The result is something like an eighth world wonder.

15 income *Einkommen*
16 oil terminal *Endstation einer Ölleitung*
17 earthquake *Erdbeben*
18 protection *Schutz*
19 wilderness *Wildnis*
20 concern *Besorgnis*
21 migration *Wanderung*

M 3 The Trans-Alaska Pipeline

M 4 The sad record of the Exxon Valdez

- 10 million birds lost.
- Plankton, sea otters, seals and whales died.
- Caribou ate oil-covered seaweed.
- £843 million spent by Exxon on the clean-up.
- Long-term effects irreparable?

The pipeline has been laid partly underground and partly above the surface. Underground, the oil must be kept warm to keep it flowing, but the soil[22] around it must be kept cold to stop the pipeline sinking and breaking. Expansion[23] of metal in warm summer temperatures and contraction[24] in winter would make the pipeline break and the oil leak out. In order to prevent[25] this, the pipeline runs on sliding shoes which allow movement.
In spite of such precautions[26], accidents still do happen and the transportation of oil is a risky business. We often hear in the news that yet another tanker has run onto the rocks in some stormy part of the world. On 24th March 1989, the oil tanker Exxon Valdez ran aground[27] in Prince William Sound, Alaska, spilling[28] 11 million gallons of crude oil and polluting 800 miles of beautiful coastline rich in marine life.

1 Find a map of Alaska in your atlas and look up the places in the text, the route of the pipeline, and the mountains and rivers that had to be crossed.
2 Explain what happens to pipelines laid in winter when the ground thaws in summer. What are the consequences for the surrounding area?
3 A large oil company has found new oil reserves in the north of Alaska. Collect ideas for a class discussion entitled 'Economy versus Ecology'. Choose representatives[29] from the oil company, the government of Alaska, the Inuit population, Greenpeace and the Alaskan National Wildlife Organization.

22 soil *Erde*
23 expansion *Ausdehnung*
24 contraction *das Zusammenziehen*
25 prevent sth. *etwas verhindern*
26 precaution *Vorsichtsmaßnahme*
27 run aground *auf Grund laufen*
28 spill sth. *etwas verschütten*
29 representative *Vertreter/-in*

Revision and further exercises

Polar Zones

Key terms

South Pole, Antarctica, Cold Pole, inland ice, shelf ice, pack ice, Arctic, Arctic Circle, iceberg, North Pole, tundra, permafrost, Inuit, Greenland, Gulf Stream, Siberia, coniferous/deciduous forest, taiga, Gazprom, Lukoil, Trans-Alaska Pipeline

Activate your English

Expressing opinions *(p. 5)*
Discussing preparations *(p. 5)*
Asking questions in an interview *(p. 9)*
Describing geographical locations and features *(p. 11)*
Describing thematic maps *(p. 13)*

M 1 Profile (cross-section) of Antarctica

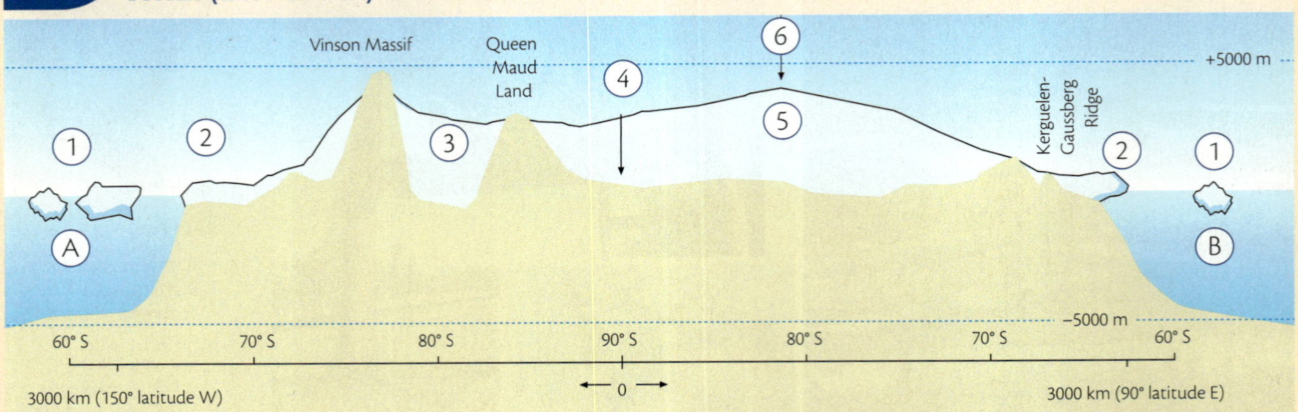

M 2 Profile (cross-section) of the Arctic

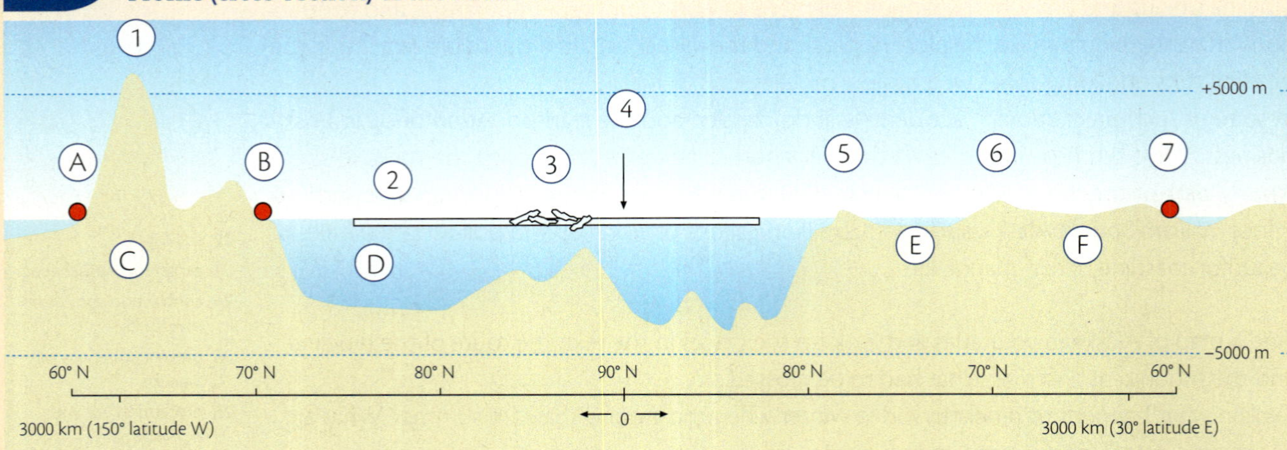

M 3 Route of the Trans-Siberian Railway

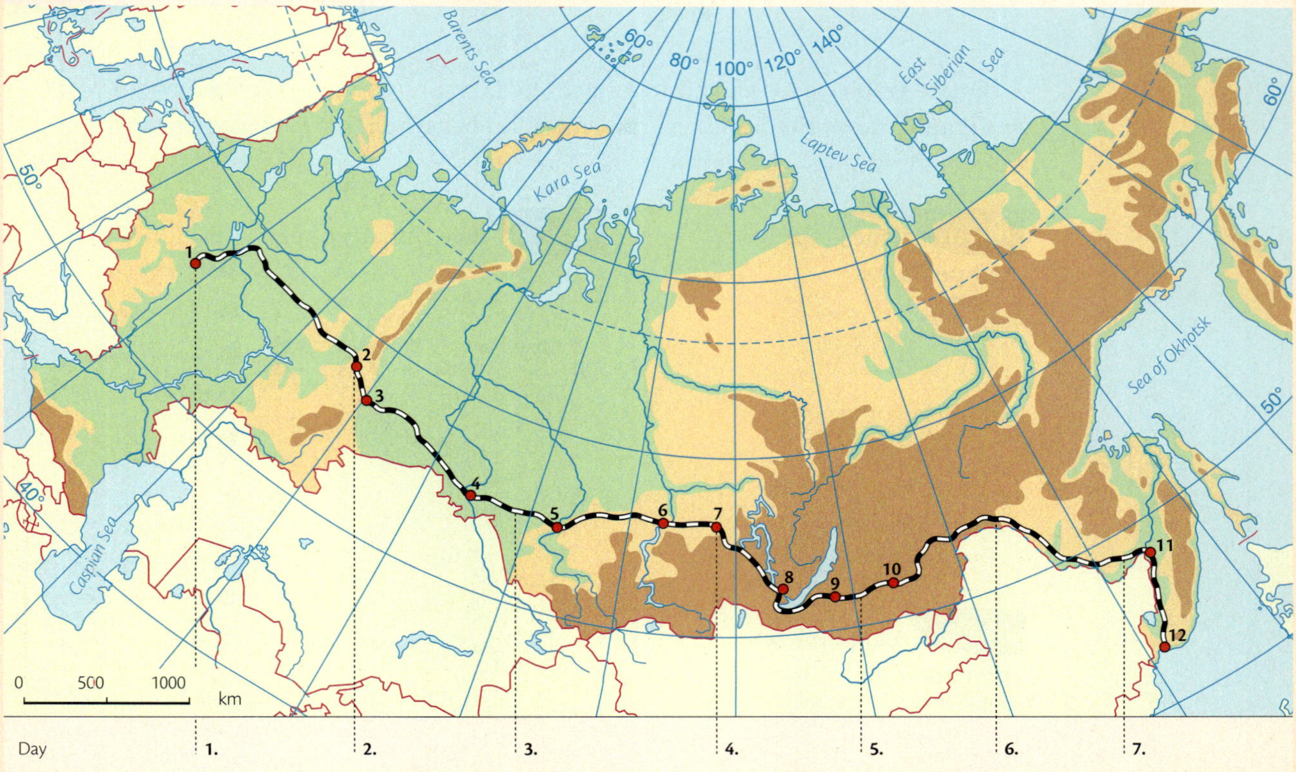

M 4 From the timetable of the Trans-Siberian Railway

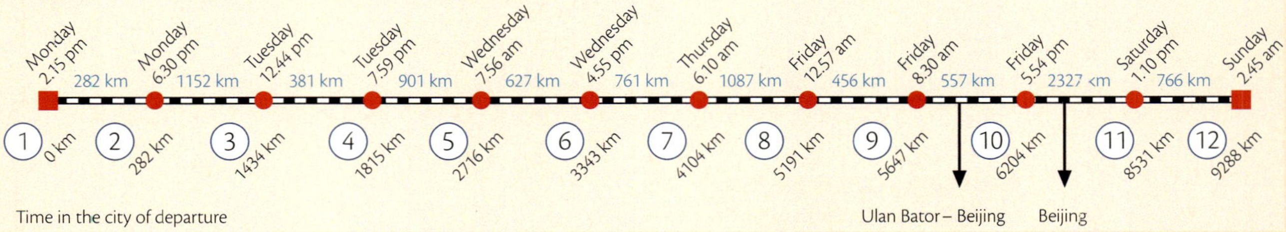

M 5 Flooding due to a rise in sea level

1. Complete the missing terms in M 1 and M 2. Write a short definition of each of the following terms: iceberg, shelf ice, inland ice, pack ice.
2. Make a quiz: Write down ten questions and answers about Antarctica and the Arctic on cards. Then get into small groups and quiz each other.
3. Look at M 3 and M 4, which give you information on the Trans-Siberian Railway. Using your atlas, name the cities numbered 1–12 along the route.
4. You have been on a journey on the Trans-Siberian Railway. Write a report. In M 3 you can see how far you travelled each day. Include the different vegetation zones the train passed through (see p. 12).
5. The map in M 5 shows how global warming could affect sea levels in Northern Germany. Describe the consequences of a rise in sea level of 5 m and 10 m.
6. Explain the reasons for the rise in global temperatures. What can we do to prevent it?

1 Reading and plotting climate graphs[1]

A **climate graph** is a special and important kind of graph. It is made up of two graphs in one – a **line graph**[2] showing temperatures (see M 1) and a **bar graph**[3] showing precipitation[4] (see M 2). Colours are added to give extra information.

M 1 Line graph for temperatures

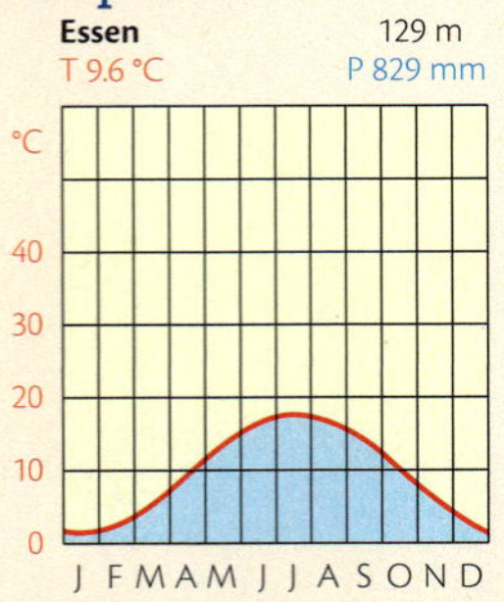

M 2 Bar graph for precipitation

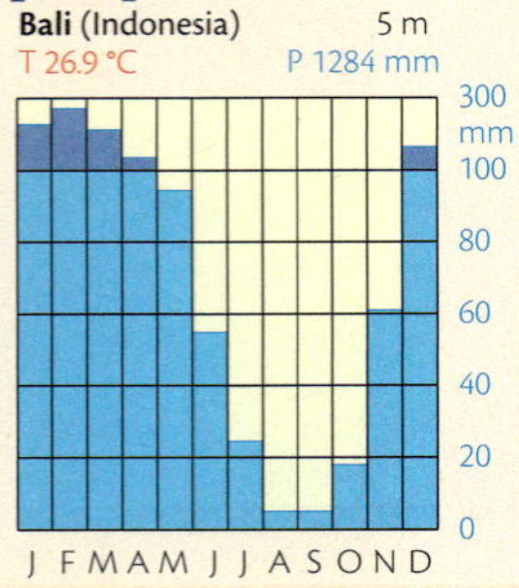

M 3 The climate graph

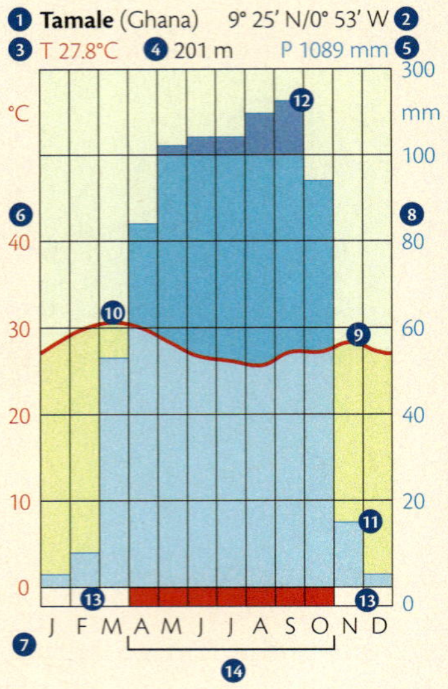

1. Climate station
2. Latitude[5]/longitude[6]
3. Mean annual temperature
4. Elevation[7]
5. Total amount of precipitation
6. Temperatures given in °C (in steps of 10°)
7. Months of the year
8. Precipitation given in millimetres (in steps of 20 mm; shortened scale above 100 mm)
9. Line graph showing temperatures
10. Maximum temperature
11. Bar graph showing precipitation
12. Maximum precipitation
13. Arid months
14. Humid months

Activate your English

Talking about a climate graph

The altitude[8] of ... is ...

Temperatures rise/increase/fall/decrease ... steadily/quickly/rapidly/slowly ... in the months of ... Temperatures reach their highest/lowest in ...

The range[9] of temperatures / difference between the highest and lowest temperature is ...

There is no / only a slight[10] / a distinct variation in temperature in ...

Temperatures are below/above/over/around ... °C for ... months of the year ... / are low / constantly high throughout the year.

Temperatures never/seldom fall below / rise above 0 °C (freezing point[11]) in ...

The mean[12]/average annual temperature is ...

The maximum/minimum precipitation is in ...

There is a rainy/dry season[13] in ...

This area/region has the least/most rain in ...

It is extremely dry/wet in ...

The climate is arid[14]/humid[15] in ...

The total rainfall/precipitation is ...

The climate of ... is extreme/mild/temperate[16] ...

The growing season[17] (see Key facts, p. 19) is in the months of / from ... to ...

1 climate graph *Klimadiagramm*
2 line graph *Liniendiagramm*
3 bar graph *Balkendiagramm, Säulendiagramm*
4 precipitation *Niederschlag*
5 (line of) latitude *Breitenkreis*
6 (line of) longitude *Längengrad*
7 elevation *Höhenangabe*
8 altitude *Höhenlage*
9 range *(hier) Schwankungsbereich*
10 slight *gering*
11 freezing point *Gefrierpunkt*
12 mean *mittlere*
13 rainy/dry season *Regen-/Trockenzeit*
14 arid *arid, trocken*
15 humid *humid, feucht*
16 temperate *gemäßigt*
17 growing season *Wachstumszeit*

1 What do each of the colours represent[18]?
2 Why are temperatures always shown as a curve[19] and precipitation in columns[20]?
3 What information is given in the heading?
4 What do you notice about the scale of the intervals on both vertical axes[21] in M 3?
5 Where is the zero on both vertical axes of the climate graph in M 3?

Plotting a climate graph

1 Think of a suitable[22] heading for your climate graph.
2 Before you begin, look at the data and decide on the scale of the vertical axis. Remember:
- to check if there are any temperatures below zero in your list;
- that the intervals for precipitation are twice as large as those for temperatures.

3 Label each axis. Write the letters for the months of the year in the middle of each column along the horizontal axis, the temperature measurements in °C along the vertical axis on the left and the precipitation measurements in mm along the vertical axis on the right.
4 Plot your figures for mean (average) monthly temperatures in °C by drawing a dot in the middle of each column.
5 Join the dots in a flowing curve.
6 Shade[23] the columns for precipitation in mm with a blue pencil. Do not draw stripes or dots.
7 Do not forget to add the mean temperature in °C and the total precipitation in mm at the top.

M 4 Climate graph of Karlsruhe

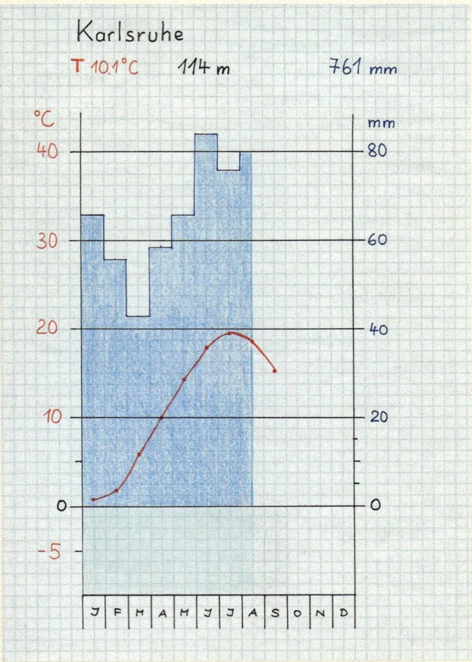

8 Now it is your turn to plot your own graphs using the following figures:

M 5 Table of climate data

51° 24' N / 6° 58' E	J	F	M	A	M	J	J	A	S	O	N	D	Year
Average temperature in °C	1.5	1.9	5.3	8.9	13.1	16.0	17.5	17.3	14.6	10.0	5.8	2.8	9.6
Total precipitation in mm	73	63	47	61	63	75	86	90	66	67	72	66	829

81° 48' N / 57° 57' E	J	F	M	A	M	J	J	A	S	O	N	D	Year
Average temperature in °C	-20.0	-20.6	-23.3	-18.3	-8.9	-1.7	0.6	0.0	-5.0	-10.6	-16.1	-18.9	-11.9
Total precipitation in mm	5	3	3	3	5	10	20	15	18	8	5	3	98

9 Find out what climate station and climate zone each of the graphs belongs to.
10 Choose one of the graphs and describe it in detail. For language help see **Activate Your English**, p. 18.
11 How long does the growing season last in each of your graphs?

Key facts

Growing season
In the growing season temperatures are above 5 °C and precipitation is greater than evaporation.

18 represent sth. *etwas darstellen, zeigen*
19 curve *Kurve*
20 column *Spalte*
21 axis (pl axes) *Achse*
22 suitable *angemessen*
23 shade sth. *etwas schraffieren*

2 Working with maps

Topographic maps[1]

M 1 Topographic map of the western USA with California

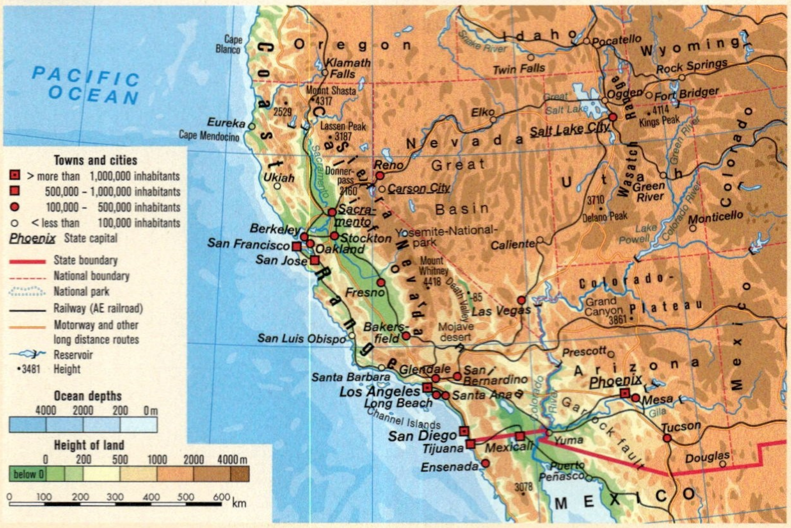

A **topographic** or **physical map** shows the surface features[2] of a landscape[3], such as mountains, lowlands[4], rivers and lakes, cities and state boundaries[5], as well as roads and railways. A legend or key[6] to the map gives more detailed information. The scale of the map is important to give some idea of the size of the area. North is always at the top of the map unless stated otherwise[7].

Activate your English

Talking about a topographic map

In the north/south/east/west/centre you can see ...
The area shown in the north/north-west ... of ... is mountainous[8]/flat[9]
The ... mountain range[10] runs[11] from ... to ... / parallel to ... /
... is situated[12] in the north/south/east/west
The River ... flows from north-east to south-west
The area ... is about/around ... m above/below sea level
Larger towns/cities are located ... in the valleys[13] / along the coast
... is densely[14]/sparsely[15] populated
Roads and railways connect/link ... (town/city) with ... (town/city) / run along the valleys
There are several large lakes/reservoirs in/along ...

1 What colours are used in topographic maps for mountains, lowlands, rivers and lakes, towns and cities?
2 What additional information is given in the key for this map?
3 How far is it from Los Angeles to San Francisco?
4 Find and name the highest mountain, the longest river, the deepest valley, two national parks and three cities with more than a million inhabitants.
5 In which direction does the Colorado River flow?
6 Write a text explaining the topographic map of California. The **Activate your English** box above and the checklist at the bottom of p. 21 will help you.

1 topographic map *topographische Karte*
2 feature *Eigenschaft*
3 landscape *Landschaft*
4 lowlands *Tiefland*
5 state boundary *Staatsgrenze*
6 key *Legende*
7 unless stated otherwise *wenn nicht anders vorgegeben*
8 mountainous *gebirgig*
9 flat *flach*
10 mountain range *Bergkette*
11 run from (place) to (place) *verlaufen von (Ort) nach (Ort)*
12 be situated somewhere *sich irgendwo befinden*
13 valley *Tal*
14 densely populated *dicht besiedelt*
15 sparsely populated *dünn besiedelt*

Thematic maps[16]

The two maps in M 2 also show California but the colours and themes are different from those in a topographic map. We call such maps **thematic maps** because they allow us to analyse a specific theme or topic in an area, which can range[17] from agriculture[18] or population[19] studies (e.g. growth or distribution[20]), to industry or nature (e.g. vegetation or climate). This type of map also has a heading, a scale and a key, which give us more detailed information about the theme of the map.

M 2 **Thematic maps of California**

1. Precipitation

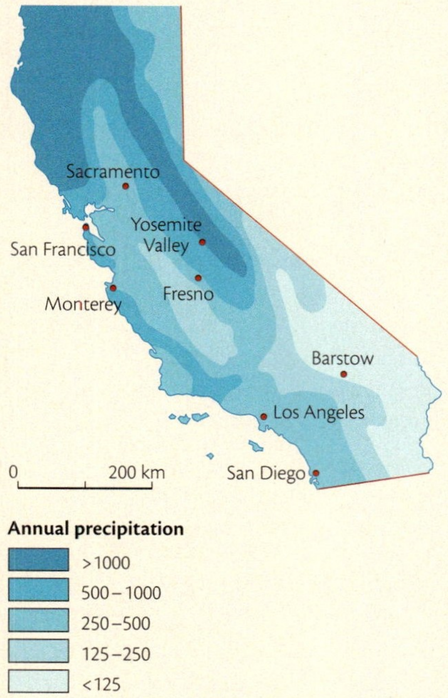

2. Population

16 thematic map *thematische Karte*
17 range from sth. to sth. *von etwas bis etwas reichen*
18 agriculture *Landwirtschaft*
19 population *Bevölkerung*
20 distribution *(hier) Bevölkerungsdichte, -verbreitung*
21 shade *Farbton*
22 even/uneven *gleichmäßig/ungleichmäßig*
23 inhabitant *Einwohner*
24 compared with sth. *im Vergleich zu etwas*
25 boundary *Grenze*
26 dominant *vorherrschend*

7 What information do these thematic maps show?
8 What do the different colours and shades[21] of colour represent?
9 What details shown in the topographic map in M 1 are missing here?
10 Choose one of the thematic maps in M 2. Make a simple sketch of it in your exercise book and write a short explanation about it. The **Activate your English** box on the right and the checklist below will help you.

Activate your English

Talking about a thematic map

The map shows … / gives information on/about …
In the north/south/east/west/centre of the map you can see …
It shows the even/uneven[22] distribution of
… more/less than … /between … and … inhabitants[23]/per square km …
There is a concentration of … in … / The population is more/less concentrated in … than in … / Very few people live in …
Precipitation is higher/lower in … than in …
Compared with[24] …, … has a high/low annual precipitation of …

Checklist for reading topographic and thematic maps

1 What is the heading of the map?
2 What scale is used in the map and how big is the area shown in reality?
3 What country/continent is the area part of and what boundaries[25] are shown?
4 What towns and cities are shown in the map? Where are they situated and how densely populated are they?
5 What are the most dominant[26] features of the landscape shown, e.g. mountain ranges, valleys, lowlands, rivers, lakes, etc.? Name some examples.
6 What other important information does the map show?

English-German vocabulary

A
above sea level *über dem Meeresspiegel*
account *Bericht, Beschreibung*
agriculture *Landwirtschaft*
altitude *Höhenlage*
aluminium ore [ɔː] *Aluminiumerz*
animal fur *Tierpelz*
annual ['ænjuəl] *jährlich*
Antarctic Circle *südlicher Polarkreis*
Arctic Circle *nördlicher Polarkreis*
arid *trocken*
axis (pl axes) *Achse*

B
Baltic Sea *Ostsee*
bar graph *Balkendiagramm, Säulendiagramm*
be entitled to sth. *auf etwas Anrecht haben*
be situated somewhere *sich irgendwo befinden*
bog *Sumpf*
boundary *Grenze*

C
calve sth. [kɑːv ☆ kæv] *etwas kälben; (hier) etwas abbrechen*
chase sb./sth. away *jdn./etwas wegjagen*
climate graph *Klimadiagramm*
coal *Kohle*
coast *Küste*
Cold Pole *Kältepol*
column ['kɒləm ☆ 'kɑːləm] *Spalte*
communication route *Verbindungsweg*
compared with sth. *im Vergleich zu etwas*
concern [- '-] *Besorgnis*
coniferous forest [kə'nɪfərəs] *Nadelwald*
conservation *Erhaltung (der Umwelt)*
construction *Anlagenbau*
contraction *das Zusammenziehen*
cope with sth. *mit etwas fertig werden*
copper *Kupfer*
cover sth. *etwas bedecken*
crevasse [krə'væs] *Gletscherspalte*
crude oil *Rohöl*
curve *Kurve*

D
deciduous forest [dɪ'sɪdʒuəs] *Laubwald*
dense *dicht*
densely populated *dicht besiedelt*
depot ['depəʊ ☆ -oʊ] *Lager*
depth *Tiefe*
develop sth. *etwas entwickeln*
distribution *(hier) Bevölkerungsdichte, -verbreitung; Verteilung*
dominant ['dɒmɪnənt ☆ 'dɑːm-] *vorherrschend*
drain away *wegsickern, ablaufen*
drift *treiben*

E
earthquake *Erdbeben*
ecological *ökologisch, Umwelt-*
ecological disaster *ökologisches Desaster, Umweltkatastrophe*
elevation *Höhenangabe*
equipment *Ausrüstung*
estimated *geschätzt*
European Union *Europäische Union*
even *gleichmäßig*
expansion *Ausdehnung*
explorer *Entdecker*
extensive flooding ['flʌdɪŋ] *weitläufige Überflutung*

F
fall into decline *verfallen, rückgängig werden*
feature ['fiːtʃə] *Eigenschaft*
fjord ['fjɔːd] *Fjord*
flat *flach*
freezing point *Gefrierpunkt*
fresh water *Süßwasser*
fringe [frɪndʒ] *Rand*
fur *Pelz*

G
glacier ['glæsiə ☆ 'gleɪʃər] *Gletscher*
government *Regierung*
growing season *Wachstumsperiode*
Gulf Stream *Golfstrom*

H
habitable *bewohnbar*
harness *Geschirr*
harsh *rau*
heather *Heidekraut*
hide *Haut*
humid ['hjuːmɪd] *humid, feucht*

I
ice sheet *Eisdecke*
income *Einkommen*
inhabitant *Einwohner*
inland ice *Inlandeis*
inlet *Meeresarm*
interior *Landesinneres*
Inuit ['ɪnjuɪt] *sing. Inuk*
iron ore *Eisenerz*

K
key *Legende (Schlüssel)*
kmh = kilometres per hour *Kilometer pro Stunde*

L
landscape *Landschaft*
layer *Schicht*
leak *lecken, auslaufen*
lichen ['laɪkən] *Flechte*
line graph *Liniendiagramm*
(line of) latitude *Breitenkreis*
(line of) longitude ['lɒŋgɪtjuːd ☆ 'lɑːn-dʒətuːd] *Längengrad*
lowlands *Tiefland*

M
marine life *Meerestiere*
marshy *sumpfig*
mean *mittlere*
mean monthly temperature *mittlere Monatstemperatur*
Mediterranean Sea [ˌmedɪtə'reɪniən] *Mittelmeer*
migration *Wanderung*
mining *Bergbau*
moss *Moos*
mountain range *Bergkette*
mountainous *gebirgig*
musk ox(en) *Moschusochse(n)*

N
natural gas reserves *Erdgasvorräte*
neglect *Vernachlässigung*
nickel *Nickel*
nickel smelter *Nickelverhüttungsanlage*
non-ferrous metal [ˌnɒn'ferəs ☆ ˌnɑːn-] *Buntmetall*
North Atlantic Drift *Golfstrom*
North Pole *Nordpol*
nutrition [nju'trɪʃn] *Ernährung*

O
ocean current *Meeresströmung*
off the hook *fertig genäht*
oil terminal *Endstation einer Ölleitung*

P
pack ice *Packeis*
peak *Bergspitze; Höhepunkt*
pension ['penʃən] *Rente*
peoples *Völker*
periodic drought [ˌpɪəri'ɒdɪk ☆ -'ɑːdɪk] *wiederkehrende Dürreperioden*
permafrost *Dauerfrostboden*
plateau ['plætəʊ ☆ plæ'toʊ] *Hochebene*
platinum *Platin*
plot a graph *ein Diagramm zeichnen*
polar bear *Eisbär*
political prisoner *politische(r) Gefangene(r)*
polluted *verschmutzt*
polyester *Polyester*

population *Bevölkerung*
precaution *Vorsichtsmaßnahme*
precipitation [prɪˌsɪpɪ'teɪʃn] *Niederschlag*
prevent sth. *etwas verhindern*
protection *Schutz*
prove to be *sich beweisen als*
provide sth. *etwas anbieten, stellen; etwas herstellen*

R

rainy/dry season *Regen-/Trockenzeit*
range *Bereich; Schwankungsbereich*
range from sth. to sth. *von etwas bis etwas reichen*
raw meat [rɔː ☆ rɑː] *rohes Fleisch*
represent sth. *etwas darstellen, zeigen*
representative *Vertreter/-in*
resource [rɪ'zɔːs ☆ 'riːsɔːrs] *Bodenschatz*
rifle *Gewehr*
rise *steigen*
run *verlaufen*
run aground *auf Grund laufen*
rust *rosten*

S

scar *Narbe*
seal *Robbe*
semi-desert *Halbwüste*
settlement *Siedlung*
sew sth. [səʊ ☆ soʊ] *etwas nähen*
shade *Farbton*
shade sth. *etwas schraffieren*
shelf ice *Eisschelf*
shelter *Schutz, Obdach*
shore *Küste, Ufer*
shrub *Strauch*
sledge [sledʒ] *Schlitten*
slide *gleiten, rutschen*
slight [slaɪt] *gering*
smelting *Verhüttung*
soggy *matschig*
soil *Boden, Erde*
South Pole *Südpol*
sparsely populated *dünn besiedelt*
spill sth. *etwas verschütten*
state boundary *Staatsgrenze*
stove *Herd, Kochstelle*
subsidy ['sʌbsədi] *Subvention*
suitable *angemessen*
supplement *Zuschuss*
supply *Vorrat*
support *Unterstützung*
surface ['sɜːfɪs] *Oberfläche*
surrounded by sth. *umgeben von etwas*

T

tamed caribou [teɪmd 'kærɪbuː] *gezähmtes Karibou*
temperate ['tempərət] *gemäßigt*
thaw [θɔː ☆ θɑː] *auftauen*
thematic map *thematische Karte*
tide *Gezeitenstrom*
tool *Werkzeug*
tooth decay *Zahnfäule*
topographic map *topographische Karte*
trading *Handel*
train oil *Lebertran*
trawler ['trɔːlə ☆ 'trɑː-] *Fischerboot*
treasure trove *Schatz, Fundgrube*
treeless *baumlos*
tundra *Tundra*

U

unemployment benefits *Arbeitslosengelder*
uneven *ungleichmäßig*
unless stated otherwise *wenn nicht anders vorgegeben*
uranium [ju'reɪniəm] *Uran*

V

valley *Tal*
valuable minerals ['væljuəbl] *wertvolle Mineralien*
vegetation *Vegetation (pflanzlicher Bewuchs)*

W

wage *Lohn*
waste *Ödland*
weapon *Waffe*
well-stocked *gut gefüllt, ausgestattet*
whale *Wal*
wilderness *Wildnis*

Abbreviations and symbols

abbr	abbreviation
AE	(mainly or exclusively) American English
BE	(mainly or exclusively) British English
C	Celsius, Centigrade
cf.	(Latin) *confer* = compare
cm	centimetre(s)
E	east
e.g.	(Latin) *exempli gratia* = for example
etc.	(Latin) *et cetera* = and so on
km	kilometre(s)
l., ll.	line, lines
m	metre(s)
mm	millimetre(s)
N	north
P	precipitation
p., pp.	page, pages
pl	plural
S	south
sb.	somebody
sing	singular
sth.	something
T	temperature
W	west
☆	American pronunciation follows
°	degree(s)
→	see Appendix

Regular differences in American pronunciation – as for instance the [r] after vowels – have not been given.

Geographical dictionary

(Analysing, discussing and arguing)

A
Achse axis (pl axes)
arid arid, dry
Ausrüstung equipment

B
Balkendiagramm bar graph, bar chart
etwas bedecken cover sth.
sich irgendwo befinden be located/situated somewhere
Bergbau mining
Bergkette mountain range
Bevölkerung population
bewohnbar habitable
Boden soil, ground
Bodenschätze resources
Breitenkreis line of latitude

D
etwas darstellen illustrate sth.
dicht dense

E
Eigenschaft feature
Einwohner inhabitant
etwas entwickeln develop sth.
Erde soil, earth
Erdgasvorräte natural gas reserves

F
feucht moist, humid
fließen flow

G
gebirgig mountainous
Gefrierpunkt freezing point
gering slight
gleichmäßig even
gleiten slide
Golfstrom Gulf Stream, North Atlantic Drift
Grenze boundary, border

H
Handel trade, trading
Hochebene plateau
Höhenangabe elevation
Höhenlage altitude

J
jährlich annual

K
Klimadiagramm climate graph
Küste shore, coast

L
Landesinneres interior
Landschaft landscape
Landwirtschaft agriculture
Längengrad line of longitude
Liniendiagramm line graph

M
Meeresspiegel sea level
Meeresströmung ocean current
Meerestiere marine life/animals
Mittelmeer Mediterranean Sea
mittlere mean

N
Niederschlag precipitation
nördlicher Polarkreis Arctic Circle
Nordpol North Pole

O
Oberfläche surface
Ostsee Baltic Sea
ökologisch ecological

P
Packeis pack ice

R
rainy/dry season Regen-/Trockenzeit
rau harsh
Regierung government
rutschen slide, slip

S
Säulendiagramm bar graph, bar chart
Schicht layer
Schutz protection
Siedlung settlement
Spalte column
Staatsgrenze state boundary
steigen increase, rise
Subvention subsidy
südlicher Polarkreis Antarctic Circle
Südpol South Pole
Süßwasser fresh water
sumpfig marshy

T
Tiefe depth
trocken arid, dry

U
Umwelt- ecological
ungleichmäßig uneven

V
im Vergleich zu etwas compared to/with sth.
etwas verhindern prevent sth.
verschmutzt polluted
Verteilung distribution
Völker peoples
Vorrat supply

W
Wachstumsperiode growing season
Wanderung migration
wegsickern drain away
Wildnis wilderness

Z
etwas zeigen indicate sth.
zunehmen increase

Acknowledgements

Titelbild: Cornelsen Verlag, Berlin; 4 (M 2) bpk; 4 (M 3) CORBIS (Bettmann); 4 (M 5) NASA Visible Earth; 5 Archiv Reinhold Messner, Juval (l); 6 Cornelsen Verlag, Berlin; 7 Cornelsen Verlag, Berlin; 8 (M 2) Cornelsen Verlag, Berlin; 8 (M 4) dpa (Polfoto Niels Hougaard); 9 (M 6) Cornelsen Verlag, Berlin; 10 (M 1) Cornelsen Verlag, Berlin; 10 (M 2) Konopka, Schwerte; 11 (M 4) Kürtz; 12 (M 2 o.) Nowosti, Berlin; 12 (M 2 M. + u.) Cornelsen Verlag, Berlin; 12 (M 4 l.) Gerd Kutschke, Berlin ;12 (M 4 r.) Nowosti, Berlin; 13 (M 5) WOSTOK, Berlin; 14 (M 2 o.) images.de (Foursa), Berlin; 14 (M 2 M.) WOSTOK, Köln; 14 (M 2 u.) Nowosti, Berlin; 15 (M 3) Cornelsen Verlag, Berlin; 15 (M 4) Dietmar Falk, Berlin